Patenting Stem Cell Technologies
Making a Claim

Colloquium
Digital Library of Life Sciences

This e-book is a copyrighted work in the Colloquium Digital Library—an innovative collection of time saving references and tools for researchers and students who want to quickly get up to speed in a new area or fundamental biomedical/life sciences topic. Each PDF e-book in the collection is an in-depth overview of a fast-moving or fundamental area of research, authored by a prominent contributor to the field. We call these e-books *Lectures* because they are intended for a broad, diverse audience of life scientists, in the spirit of a plenary lecture delivered by a keynote speaker or visiting professor. Individual e-books are published as contributions to a particular thematic **series**, each covering a different subject area and managed by its own prestigious editor, who oversees topic and author selection as well as scientific review. Readers are invited to see highlights of fields other than their own, keep up with advances in various disciplines, and refresh their understanding of core concepts in cell & molecular biology.

For the full list of published and forthcoming Lectures, please visit the Colloquium homepage: www.morganclaypool.com/page/lifesci

Access to the Colloquium Digital Library is available by institutional license. Please e-mail info@morganclaypool.com for more information.

Morgan & Claypool Life Sciences is a signatory to the STM Permission Guidelines. All figures used with permission.

Colloquium Series on Stem Cell Biology

Editor

Wenbin Deng, Ph.D., *Cell Biology and Human Anatomy, Institute for Pediatric Regenerative Medicine, School of Medicine, University of California, Davis*

This Series is interested in covering the fundamental mechanisms of stem cell pluripotency and differentiation, as well as strategies for translating fundamental developmental insights into discovery of new therapies. The emphasis is on the roles and potential advantages of stem cells in developing, sustaining, and restoring tissue after injury or disease. Some of the topics covered include the signaling mechanisms of development and disease; the fundamentals of stem cell growth and differentiation; the utilities of adult (somatic) stem cells, induced pluripotent stem (iPS) cells and human embryonic stem (ES) cells for disease modeling and drug discovery; and finally the prospects for applying the unique aspects of stem cells for regenerative medicine. We hope this Series will provide the most accessible and current discussions of the key points and concepts in the field, and that students and researchers all over the world will find these in-depth reviews to be useful.

For a list of published and forthcoming titles:
http://www.morganclaypool.com/page/scb

Copyright © 2013 by Morgan & Claypool

Patenting Stem Cell Technologies:
Making a Claim
Antoinette F. Konski
www.morganclaypool.com

ISBN: 9781615046225   paperback

ISBN: 9781615046232   ebook

DOI: 10.4199/C00078ED1V01Y201303SCB004

A Publication in the

COLLOQUIUM SERIES ON STEM CELL BIOLOGY

Lecture #4

Series Editor: Wenbin Deng

Series ISSN
ISSN 2168-3972   print
ISSN 2168-3980   electronic

Patenting Stem Cell Technologies

Making a Claim

Antoinette F. Konski
Foley & Lardner LLP

COLLOQUIUM SERIES ON STEM CELL BIOLOGY #4

MORGAN & CLAYPOOL PUBLISHERS

ABSTRACT

Are stem cells patentable? What is the patenting process? What rights does a patent provide? Why should I patent?

Applying for and obtaining a patent is a process that can be unpredictable and intimidating, although it does not necessarily need to be. Novice and experienced inventors often have questions regarding patenting and the patenting process. This e-book is provided to answer many questions regarding the patenting process before the United States Patent and Trademark Office ("USPTO"). It also generally describes the technologies typically patented in connection with regenerative medicine.

This e-book is provided for informational purposes only and should not replace legal advice, which is necessary to anticipate and address the nuances of the patenting process. In addition, there are issues that should be considered and addressed when considering patenting isolated stem cells and associated technologies—such as the process for obtaining patent rights outside the United States, post-grant procedures for challenging patents, non-patent protection of intellectual property, and enforcement of patents through litigation—which are beyond the scope of this chapter.

KEYWORDS

patents, patenting process, patentability, patent claims, patent applications, patent examination process, intellectual property, USPTO, regenerative medicine, stem cells, stem cell technologies

Contents

CHAPTER 1

The U.S. Patent System

1.1   THE LEAHY-SMITH AMERICA INVENTS ACT ("AIA")

The information provided in this e-book is in accordance with the patenting process under the Leahy-Smith America Invents Act ("AIA").[1] This legislation was signed into law by President Barack Obama on September 16, 2011.[2] The law is the most significant change to the U.S. patent system since 1952.

One significant impact of the AIA is that the U.S. patent system went from awarding a patent to the inventor who was "first-to-invent" to the inventor who is the "first-to-file" the patent application. As such, it eliminates interference procedures that determine who, among two or more inventors, was the first-to-invent the claimed technology. The "first-to-file" provisions of the AIA are applied to any application that has an effective filing date of March 16, 2013 or later.[3] Additional changes to the U.S. patenting process include the adoption of: derivation proceedings, a new micro entity status, post-grant review of applications and an expanded defense of prior user rights.[4]

---

[1] An excellent reference for the patenting process prior to the enactment of the AIA is "What Every Chemist Should Know About Patents," written and edited by Le-Nhung McLeland for the ACS Joint Board-Council Committee on Patents and Related Matters, available at //portal.acs.org /portal/fileFetch/C/WPCP_006903/pdf/WPCP_006903.pdf.

[2] Leahy-Smith America Invents Act ("AIA"), Pub. L. No. 112-29, 125 Stat. 284 (2011).

[3] For a detailed analysis of the patenting process under the AIA as compared to the prior laws and rules, see American Invents Act: Law & Analysis, 2013 Edition, Wolters Kluwer, CCH Incorporated (2013).

[4] As noted herein, these additional nuances of the patenting process are beyond the scope of this e-book. Inventors and those involved with the patenting process are advised to consult a professional.

1.2   THE LEGAL BASIS FOR U.S. PATENT RIGHTS

Article I, section 8 of the U.S. Constitution is the basis for the grant of patent rights in the United States. It states that Congress "shall have power . . . to promote the progress of science and useful arts, by securing for limited times to . . . inventors the exclusive right to their . . . discoveries."

Patents in the U.S. are governed by the Patent Act (Title 35 of the U.S. Code or "35 U.S.C."), which established the United States Patent and Trademark Office. Title 35 provides that "whoever invents or discovers any new and useful process, machine, manufacture, or composition of matter, or any new and useful improvement thereof, may obtain a patent therefor, subject to the conditions and requirements of this title." (35 U.S.C. § 101).

Title 37 of the Code of Federal Regulations ("37 C.F.R.") is the codification of the rules published in the Federal Register by the executive departments and the USPTO. These rules establish the day-to-day procedures and processes governing the patent application process.

The USPTO publishes a guide for patent examiners that describes all the laws (Title 35 of the U.S. Code) and regulations (Title 37 of the U.S. Code of Federal Regulations) and their application to the patent examination process. This "Manual of Patent Examining Procedure" ("MPEP") is a resource not only for patent examiners but also for anyone involved in the patenting process, e.g., patent attorneys, patent agents and inventors. The complete MPEP is available on the USPTO web page at: uspto.gov/web/offices/pac/mpep/index.html.

1.3   WHAT IS A PATENT?

A patent is a property right issued by the government to an inventor or a group of inventors when more than one individual jointly makes the invention. For the sake of convenience in this chapter, the terms "patentee," "inventor," and "applicant" shall include the case of a singular and joint inventor(s), applicant(s) or patentee(s).[5]

A patent allows the patent holder to prevent others from making, using, importing or selling (a.k.a. "practicing") the claimed technology in the country where the patent is granted. The individual or individuals who practice the invention described in a patent claim without authorization from the patentee are said to infringe the patent claim.

Section 271 of Title 35 sets forth several theories of infringement, e.g., direct infringement, contributory infringement and inducement.[6] U.S. patent law also allows a patentee the ability to stop the importing into the United States products made by a U.S. patent granted on the process of making the product.

It is important to note that a U.S. patent does not give the patentee the right to practice the technology free from liability to any other patentee; it only grants the right to stop others who are not a patentee (or one having permission from the patentee) from practicing it. Other patents may exist that prevent the patentee from using the patented technology because the other patent rights dominate the patentee's property right.

While a patent provides the patentee the right to exclude others from practicing the claimed technology, a patent owner can choose not to enforce the patent rights and allow others to freely use it. Alternatively, the patentee can grant others a license to use the patented technology for a fee or other consideration. The license granted may be for all or a portion of the patented technology,[7]

---

[5] An inventor or an assignee (as applicant) can file a patent application.

[6] A person can be liable under 35 U.S.C. § 271 for making, using, selling, or importing into the US any patented invention, without authority, during the term of the patent. Infringement can be "direct" or "indirect." Typically, infringement under 35 U.S.C. § 271(a) is direct infringement while infringement under 35 U.S.C. §§ 271 (b) and (c) are typically termed indirect infringement. Direct infringement does not require knowledge of the patent or any intent to infringe. In contrast, indirect infringement can only arise when the accused indirect infringer has at least some knowledge and intent regarding the patent and the infringement.

[7] A patentee can license the patent for a stated field of use, exclusively or non-exclusively or for a particular geography, exclusively or non-exclusively.

and it can be exclusive to one party or non-exclusive to more than one party. An exclusive license allows only one party to practice the licensed technology while a non-exclusive license allows more than one party to practice the licensed technology. Similar to other property rights, patent rights can be transferred in whole to another by an assignment agreement.

The key difference between an assignment and license is the ownership of a patent. With a license, ownership remains with the patentee, it is not transferred to the licensee. In contrast with an assignment, ownership of the patent rights is transferred in whole to the assignee and the patentee retains no rights.[8]

In the United States, the USPTO is the granting authority for the territory of the United States, U.S. territories, and U.S. possessions.[9] If an inventor wishes to prevent others from making, using or selling in a country other than the United States, the inventor must apply for a patent in each country or region where a patent grant is desired. Thus for example, if an inventor wants the right to exclude others from making, using or selling the technology in Canada, an application for a patent must be applied for in Canada and the Canadian Patent Office grants a Canadian patent.

1.4 WHY SHOULD I PATENT?

A patent is a lawful and limited monopoly that provides a period during which a patentee can exclude others from practicing the same technology. It therefore provides a business advantage over competitors. It also provides innovators with an opportunity to create and consolidate a market position, provide a revenue stream through licensing, franchise or sale, raise capital, increase power in a business negotiation and in general, allows one the time to develop a product and gain a return on investment. Alternatively, a patent can be sold (assigned) or licensed

---

[8] A license is analogous to a property rental agreement. The renter has the right to live and use the property according to the terms of the rental contract, but title to the property remains with the title holder. An assignment is a sale of the property and title to the property is transferred to the assignee, with no rights remaining with the assignor.

[9] The USPTO provides a general review of the patenting process at uspto.gov/patents/process/index .jsp.

for a fee or other consideration such as for a licensee to practice a patent owned by another party.

A patent also supports the release of technology into the public domain. After expiration of the patent term, the technology is available to anyone without recourse or liability to the original patentee. Indeed, the use of patented technology after the expiration of patent term is the basis for the generic drug market.

1.5   PATENT TYPES

There are three types of patents issued in the United States: a) utility patents, b) design patents, and c) plant patents.

1.5.1   Utility Patents

Utility patents may be granted to anyone who invents or discovers any new and useful process, machine, article of manufacture, or composition of matter, or any new and useful improvement thereof. Most patents in the regenerative medicine field are utility patents.

1.5.1.1 Provisional vs. Nonprovisional Patents

There are two types of utility applications—provisional and nonprovisional applications. Only a nonprovisional application can mature into a patent. A provisional application cannot mature into a patent but provides a lower cost alternative to a nonprovisional application as a means to secure a filing date. Additionally, the requirements for filing a provisional application are minimal as compared to a nonprovisional application.[10]

---

[10] A provisional application for patent is a U.S. national application for patent filed in the USPTO under 35 U.S.C. § 111(b). It allows filing without a formal patent claim, oath or declaration, or any information disclosure (prior art) statement. It provides the means to The option to file a provisional patent application before the USPTO began on June 8, 1995. Prior to that date, only utility applications could be filed with the USPTO. The filing of a provisional patent application also allows the term "Patent Pending" to be applied in connection with the description of the invention. For

After the expiration of one year from the date of filing, a provisional application expires with no rights outstanding to the inventor. Thus, because a provisional application cannot mature into a patent, the provisional application must be refiled as a nonprovisional patent application within one year from the anniversary of the original filing date. If the inventor refiles the provisional application within the one year period as a nonprovisional application and complies with other legal requirements, the inventor is entitled to claim the filing date and benefit of the provisional application. In addition, the period between the filing of the provisional application and the nonprovisional application is not included in the patent term.

1.5.2   Design Patents

Design patents may be granted to anyone who invents a new, original, and ornamental design for an article of manufacture.[11]

The design for an article consists of the visual characteristics embodied in or applied to an article, for example, the configuration or shape of an article, the surface ornamentation applied to an article, or to the combination of configuration and surface ornamentation.

While utility and design patents afford legally separate protection, the utility and ornamentality of an article may not be easily separable. Articles of manufacture may possess both functional and ornamental characteristics, but they are different in terms of the scope of the patent and patent term.[12]

---

more information on the filing of and requirements of filing a provisional application, see uspto.gov/patents/resources/types/provapp.js.

[11] Chapter 1500 of the MPEP provides information on design patent protection. Because design patents are not typically pursued for technologies in regenerative medicine, they are not covered in this chapter.

[12] MPEP §1502 lists and describes several key distinctions between utility and design patents. These include for example patent term, maintenance requirements and claiming requirements.

1.5.3 Plant Patents

Plant patents may be granted to anyone who invents or discovers and asexually reproduces any distinct and new variety of plant.[13]

1.6 CRITERIA FOR PATENTABILITY IN THE UNITED STATES

As noted above and as set forth in 35 U.S.C. § 101, an invention must be new (novel), non-obvious, have utility and be directed to patent-eligible subject matter.

1.6.1 35 U.S.C. § 101—Patent-Eligible Subject Matter

35 U.S.C. § 101[14] defines four categories of inventions that are patent-eligible: processes, machines, manufactures and compositions of matter. The latter three categories define "things" or "products" while the first category defines "actions" (i.e., inventions that consist of a series of steps or acts to be performed). Abstract ideas, laws of nature and physical phenomena have been determined by the courts of the United States to be outside the scope of patentable subject matter.[15] Stem cells and other material derived from naturally occurring products are considered patent-eligible as long as they are claimed as "isolated" or "purified" products to distinguish them from the material in the natural environment.

In the context of the patent-eligibility of isolated human DNA, to what extent naturally occurring material is patent-eligible is currently under review by the U.S. Supreme Court in *Ass'n for Molecular Pathology v. Myriad Genetics, Inc.*,

---

[13] See MPEP Chapter 1600 for more information on plant patents.

[14] Section 101 of Title 35 of the U.S. Coder provides that "[w]hoever invents or discovers any new and useful process, machine, manufacture, or composition of matter, or any new and useful improvement thereof, may obtain a patent therefor, subject to the conditions and requirements of this title."

[15] MPEP § 2103, *citing Bilski v. Kappos*, 130 S. Ct. 3218, 3225 (2010) (*citing Diamond v. Chakrabarty*, 447 U.S. 303, 309 (1980)).

decision below at No. 2010-1406 (Fed. Cir. 2012). A decision is expected in 2013. Although the Supreme Court is determining whether human genes are patentable, the Supreme Court's opinion could affect the patent-eligibility of other material derived from the human body, such as stem cells isolated from a human.

The requirement of 35 U.S.C. § 101 is considered a threshold inquiry for the grant of a patent.[16] Even if an invention qualifies as a process, machine, manufacture, or composition of matter, the invention must also satisfy "the conditions and requirements of this title." The additional requirements are: the invention is **novel** (in accordance with 35 U.S.C. § 102), **non-obvious** (in accordance with 35 U.S.C. § 103), and **fully and particularly described** (in accordance with 35 U.S.C. § 112).

1.6.1.1 35 U.S.C. § 102—Novelty

Section 102 of Title 35 of the U.S. Code defines novelty, novelty-destroying activities and exceptions to novelty-destroying activities:

"Sec. 102. Conditions for patentability; novelty

(a)   Novelty; Prior Art—A person shall be entitled to a patent unless—
   1.   the claimed invention was patented, described in a printed publication, or in public use, on sale, or otherwise available to the public before the effective filing date of the claimed invention; or
   2.   the claimed invention was described in a patent issued under section 151, or in an application for patent published or deemed published under section 122(b), in which the patent or application, as the case may be, names another inventor and was effectively filed before the effective filing date of the claimed invention.

---

[16]MPEP § 2103.

(b) Exceptions—

1. DISCLOSURES MADE 1 YEAR OR LESS BEFORE THE
    EFFECTIVE FILING DATE OF THE CLAIMED INVENTION—
    A disclosure made 1 year or less before the effective filing date of a
    claimed invention shall not be prior art to the claimed invention under
    subsection (a)(1) if—

    A. the disclosure was made by the inventor or joint inventor or by another
        who obtained the subject matter disclosed directly or indirectly from
        the inventor or a joint inventor; or

    B. the subject matter disclosed had, before such disclosure, been pub-
        licly disclosed by the inventor or a joint inventor or another who
        obtained the subject matter disclosed directly or indirectly from the
        inventor or a joint inventor.

2. DISCLOSURES APPEARING IN APPLICATIONS AND
    PATENTS—A disclosure shall not be prior art to a claimed invention
    under subsection (a)(2) if—

    A. the subject matter disclosed was obtained directly or indirectly
        from the inventor or a joint inventor;

    B. the subject matter disclosed had, before such subject matter was
        effectively filed under subsection (a)(2), been publicly disclosed by
        the inventor or a joint inventor or another who obtained the subject
        matter disclosed directly or indirectly from the inventor or a joint
        inventor; or

    C. the subject matter disclosed and the claimed invention, not later
        than the effective filing date of the claimed invention, were owned
        by the same person or subject to an obligation of assignment to the
        same person.

(c)   Common ownership under joint research agreements.—

Subject matter disclosed and a claimed invention shall be deemed to have been owned by the same person or subject to an obligation of assignment to the same person in applying the provisions of subsection (b)(2)(C) if—

"(1) the subject matter disclosed was developed and the claimed invention was made by, or on behalf of, 1 or more parties to a joint research agreement that was in effect on or before the effective filing date of the claimed invention;

(2) the claimed invention was made as a result of activities undertaken within the scope of the joint research agreement; and

(3) the application for patent for the claimed invention discloses or is amended to disclose the names of the parties to the joint research agreement.

(d)   Patents and Published Applications Effective as Prior Art- For purposes of determining whether a patent or application for patent is prior art to a claimed invention under subsection (a)(2), such patent or application shall be considered to have been effectively filed, with respect to any subject matter described in the patent or application—

1.   if paragraph (2) does not apply, as of the actual filing date of the patent or the application for patent; or

2.   if the patent or application for patent is entitled to claim a right of priority under section 119, 365(a), or 365(b), or to claim the benefit of an earlier filing date under section 120, 121, or 365(c), based upon 1 or more prior filed applications for patent, as of the filing date of the earliest such application that describes the subject matter."

* * *

For an invention to be patentable, it must be "novel" or not described or disclosed to the public by the inventor more than one year before the filing date of the application (called a one-year grace period) or prior to the disclosure by a non-inventor prior to the filing date of the application. While the inventor has a one-year grace period from the date of publication or other novelty-destroying activity to file a patent application, the one-year exception does not apply to novelty-destroying disclosures made by one who is not an inventor.

A "disclosure" of the invention can be public dissemination of the invention by acts that include publication in print or electronic media, in a speech, public use of the invention, or commercialization of the invention by a sale or offer of sale of the claimed invention anywhere in the world.

If a patent application is not filed within the one-year period from the inventor's first disclosure, publication, public use or commercialization of the invention, the inventor will be barred from obtaining a patent on the invention. In addition, very few foreign countries allow for a grace period from disclosure to patent filing. Therefore, if an inventor wants to seek patent protection in foreign countries, the most prudent course of action is to file a patent application describing the invention before any disclosure of the invention.

As noted above, one of the most significant changes that the AIA made to the patenting process is the adoption of a "first-to-file" system from a "first-to-invent" system. Prior to March 16, 2013, if two patent applications claimed (or a patent and a pending application) the same invention, the patent would be issued to the inventor who was the first-to-invent the claimed technology. The determination of who was the first-to-invent the claimed invention was determined in an interference procedure conducted in the USPTO. Under the AIA, however, if two patent applications claim the same invention, the patent will be awarded[17] to the

---

[17] Provided of course the other conditions of patentability are met—novelty, non-obviousness, utility and specificity of description.

application that has the earliest filing date.[18] An exception exists when it is shown that the inventor of the first filed application derived his invention from the inventor of the subsequently filed application.

For publications or disclosures that pre-date the filing date of the application, the publication or disclosure will prevent a patent on the invention if the publication discloses every element of the claim. In addition, the publication or disclosure must describe the invention in a manner that would allow one of skill in the art to reproduce the technology.

As noted above, there is an exception to the novelty requirement if the patent-defeating acts occurred less than one year from the filing date of an inventor's application. An inventor's own disclosure of the invention, if made less than one year prior to the filing date of an application, will not be prior art against the application. If a disclosure by one other than the inventor is made prior to the filing date of an application, but the information in the disclosure was obtained directly or indirectly from the inventor, the disclosure is not prior art.

In addition, if one other than the inventor files an application prior to the inventor's application, but the subject matter of the other application was derived from an inventor's prior disclosure, the prior filed application is not prior art. However, in order for the exception to apply, the inventor must have filed his own application within one year from his own publication.

Section 102 also excludes from patent-defeating disclosures, commonly owned, prior published U.S. patent applications if they were commonly owned at the time of the effective filing date of the application. In addition, if the prior filed application and the invention were made by, or behalf of one (1) or more parties to a joint research agreement, then the prior filed application is not prior art to the

---

[18]The USPTO will continue determining the first-to-invent for applications that have filing dates that straddle March 16, 2013. An example is if a provisional application was filed prior to March 16, 2013 and new material information was added to the provisional application and it was filed as a nonprovisional on or after March 16, 2013. However, new 35 U.S.C. §§ 102 and 103 will be applied in determining novelty and non-obviousness.

second filed application. A joint research agreement is a written contract, grant or cooperative agreement entered into by two (2) or more persons or entities for the performance of experimental, development or research work in the field.[19]

1.6.1.2 35 U.S.C. § 103—Non-Obviousness

Section 103 of Title 35 of the U.S. Code defines non-obviousness:

35 U.S.C. 103 Conditions for patentability; non-obvious subject matter.

"A patent for a claimed invention may not be obtained, notwithstanding that the claimed invention is not identically disclosed as set forth in section 102, if the differences between the claimed invention and the prior art are such that the claimed invention as a whole would have been obvious before the effective filing date of the claimed invention to a person having ordinary skill in the art to which the claimed invention pertains. Patentability shall not be negated by the manner in which the invention was made."

* * *

An invention that would have been obvious to a person of ordinary skill as of the effective filing date of the application is not patentable. Where a prior disclosure does not disclose each and every element of a claimed invention, it can be combined with other references or common knowledge to defeat patentability if the information in combination would have suggested or led one to the claimed invention.[20]

---

[19] Joint research agreement is defined in 35 U.S.C. § 103(c)(3). 37 C.F.R. § 1.104(c) provides the rules to implement this exception.

[20] The question of obviousness is resolved on the basis of certain factual determinations that have been developed by the U.S. courts.

1.6.1.3 35 U.S.C. § 112—Disclosure Requirements

An invention that meets the three statutory requirements of utility, novelty, and non-obviousness nevertheless cannot be patented unless the patent application discloses the invention in a manner that describes it in sufficient detail to enable any person skilled in the art to reproduce the invention throughout the full scope of the claimed invention.

The law setting forth this requirement is found in 35 U.S.C. § 112(a):

> "The specification shall contain a written description of the invention, and of the manner and process of making and using it, in such full, clear, concise, and exact terms as to enable any person skilled in the art to which it pertains, or with which it is nearly connected, to make and use the same, and shall set forth the best mode contemplated by the inventor or join inventor of carrying out the invention."

1.7   INVENTORS, INVENTORSHIP, AND OWNERSHIP

In the United States, an inventor is the person (or joint inventor if the invention is made by more than one person) who conceived at least one element of the claims of patentable invention. Unless the person contributed to the conception of the invention, he or she is not an inventor. "Conception" is defined as the complete performance of the mental part of the inventive act. The invention must also be "reduced to practice" either by actual experimentation, or by describing the invention in a patent application in sufficient detail that one of skill in the art can reproduce the invention with a minimal amount of experimentation. While these concepts seem simple enough, the application of the doctrines to other than the most straightforward facts is difficult. Inventorship can change during the course of examination or as more facts come to light after the invention has been made or as the claim language is amended during examination.

An inventor owns an invention made in the United States until and unless the inventor transfers ownership to another by contract. Where more than one individual is an inventor, each inventor has the independent right to transfer ownership or license the application or patent without approval of the other co-inventor(s). It is stated that each co-inventor has an undivided equal interest in the application or patent. In addition, an inventor need only contribute to one element of one claim of an application or patent. Nevertheless, the co-inventor has the equal undivided interest in the whole of the application or patent.

1.8   PATENT TERMS

A patent is enforceable only after it issues. A "patent term" is the period during which a patent is enforceable against a third party. Pending patent applications are not enforceable against third parties, although certain provisional rights are available if specific criteria are met.[21]

The term of U.S. utility patents was significantly amended in 1995. For utility applications filed in the United States on or after June 8, 1995,[22] the term of a patent (other than a design patent) begins on the date the patent issues and ends on the date that is twenty years from the date on which the application for the utility patent was filed in the United States or, if the application contains a specific reference to a specific application,[23] twenty years from the filing date of the earliest of such application(s). Design patents have a term of fourteen years from the date of patent grant.[24]

All patents (other than U.S. design patents) that were in force on June 8, 1995, or that issued on an application that was filed before June 8, 1995, have a term that is the greater of the term beginning on the earliest filing date claimed

---

[21] See 35 U.S.C. § 154 for the rights and requirements for provisional rights.
[22] The term of a U.S. patent is governed by Section 532(a)(1) of the Uruguay Round Agreements Act (Pub. L. 103-465, 108 Stat. 4809 (1994)) amended 35 U.S.C. § 154.
[23] The specific applications are determined by reference to 35 U.S.C. §§ 120, 121 or 365(c).
[24] See 35 U.S.C § 173 and MPEP § 1505.

or seventeen years from date of the grant of the patent.[25] A patent granted on an international application filed before June 8, 1995, and which entered examination before the USPTO (as a national stage application under 35 U.S.C. § 371, before, on or after June 8, 1995), will have a term that is the greater of seventeen years from the date of grant or twenty years from the international filing date or any earlier filing date of a specified application.[26]

In the United States, the term of a patent may be subject to reduction by a terminal disclaimer or for delays caused by a patent applicant during the application process. In some instances, the USPTO determines that inventions claimed in two or more patent applications or issued patents, filed by the same inventor, are substantially the same invention. The USPTO will grant a patent on the second or later issued patent application to the inventor if the inventor agree to limit the patents to expire on the same date. The agreement to do so is a "terminal disclaimer."

The terms of utility and plant patents issuing on applications filed on or after May 29, 2000 also may be adjusted for delays during the patent application process. The adjustment may add to the patent term for certain delays caused by the USPTO or may subtract from the patent term for delays caused by the inventor. Plant and utility patents issuing on applications filed before June 8, 1995, which have a term that is the greater of the twenty-year term or seventeen years from patent grant, are not eligible for term extension or adjustment due to delays in processing the patent application by the United States Patent and Trademark Office.[27] Alternatively or in addition, a U.S. patent term can be extended for time lost while awaiting marketing approval from the U.S. Food and Drug Administration.[28]

---

[25] See 35 U.S.C. § 154.
[26] The specific applications are those filed under the provisions of 35 U.S.C. §§ 120, 121 or 365(c).
[27] See, in general MPEP Chapter 2700.
[28] For more information on extensions of the patent term due to the regulatory review process, see www.fda.gov/Drugs/DevelopmentApprovalProcess/SmallBusinessAssistance/default.htm.

Since the term of a design patent is not affected by the length of time prosecution takes place, there are no patent term adjustment provisions for design patents.[29]

Because the development period of stem cell technologies is long, most patents in stem cell technologies are first filed as provisional applications and then refiled as a nonprovisional application, thereby extending the patent term to up to 21 years from the date of the provisional filing.[30]

·   ·   ·   ·

---

[29] *Id.*

[30] Nonprovisional patent applications that claim priority to a provisional patent application must claim domestic priority to the provisional application under 35 U.S.C. § 119(e). The term between the date of the provisional application and the nonprovisional application is not considered in the calculation of the twenty-year term. See 35 U.S.C. § 154(a)(3) and MPEP § 2701.

CHAPTER 2

The Patenting Process

2.1   INVENTION

The patenting process begins with an invention. An invention is a composition, device, process or method that has been conceived through thought by one or more individuals (i.e., inventors) and reduced to practice. However, not all inventions are patentable. The process of determining whether an invention, described in a patent application is patentable, is the patent examination process.

2.2   REDUCTION TO PRACTICE

The sole or jointly conceived idea may or may not have been tested by an actual experiment. When the invention has been tested and proven by experiment, the invention is said to have been "actually reduced to practice." U.S. patent law also allows an inventor to file a patent application on an idea that has not yet been tested and confirmed by an actual experiment. When an invention has not yet been confirmed by an actual reduction to practice, the patent application must describe the invention in sufficient detail in the patent application such that another person having sufficient background and experience in the field of the invention (a.k.a. "one of skill in the art")[1] could reproduce the invention without an undue amount of experimentation. The USPTO may also require evidence that

---

[1] A "person having ordinary skill in the art" (a.k.a. PHOISTA) is a fictitious person who is used as a basis for comparison against which an invention or patent application is measured. The fictional person is considered to have the education, knowledge and skills of an ordinary artisan in the field of the invention.

the invention performs actually as predicted in the patent application. This can be established with experimental data presented after the filing of the application papers and during the examination process.

2.3   FILING AN APPLICATION

After an invention has been made and the decision has been made to apply for a patent, the application should be drafted to properly describe the invention. The application can be drafted and filed as a provisional utility application or as a non-provisional utility application. A provisional application should contain a patent specification that complies with 35 U.S.C. §112 (written description, enablement and best mode) and any drawing that may be required to properly describe the invention.[2] A cover sheet identifying the application as a provisional application and payment of the appropriate fee is also required. If the application is not identified as a provisional application, the USPTO will treat the application as a nonprovisional application.[3]

It may be useful to search for references or other disclosures that could impact the scope of a patentable invention or completely prevent it, by disclosing or making obvious the invention. While searching is best undertaken by a professional searcher and reviewed by the one who will be preparing the application, the USPTO provides guidance to the public on searching strategies.[4]

2.3.1   Provisional Applications

Provisional applications are not examined and cannot mature into a patent. They provide an alternative to filing an initial nonprovisional application. They will provide a filing and priority date for the inventions that are described in the provisional application in accordance with the provisions of 35 U.S.C. §112 (written

---

[2] See MPEP § 603.

[3] Id.

[4] See uspto.gov/web/offices/ac/ido/ptdl/CBT/.

description, enablement and best mode) but must be "converted" to a nonprovisional application on or before the one year anniversary date of the filing date of the provisional and the provisional must be claimed as a priority filing in the nonprovisional patent application. A provisional application automatically becomes abandoned after the one-year anniversary date.

2.3.2   Nonprovisional Applications

A nonprovisional application must also include a description of the invention that complies with 35 U.S.C. §112 (written description, enablement and best mode) and at least one claim and any drawing that may be required to properly describe the invention. Nonprovisional applications may refer to an earlier filed application, such as an earlier filed provisional application or an earlier filed nonprovisional application.[5] A properly drafted nonprovisional patent application should include at least a cover page that identifies the inventors and the title of the invention. The cover page may also cite any prior filed priority application to which priority is claimed. The assignee and attorney or agent that is representing the inventors before the USPTO also may be listed on the cover page.[6]

2.3.3   Patent Specification

The main body of the patent application is the patent specification. As determined by the patent rules, the specification of a nonprovisional application should include the following sections in order:[7]

   (1) Title of the invention, which may be accompanied by an introductory portion stating the name, citizenship, and residence of the applicant (unless included in the application data sheet).

---

[5] When an application refers to an earlier filed application, the nonprovisional application is said to "claim the benefit of" or "claim priority to" an earlier filed application. Provisional applications cannot claim the benefit of an earlier filed provisional or nonprovisional application.
[6] See MPEP § 601.
[7] See 37 C.F.R. § 1.77.

(2) Cross-reference to related applications (unless included in the application data sheet).

(3) Statement regarding federally sponsored research or development.

(4) The names of the parties to a joint research agreement.

(5) Reference to a "Sequence Listing," a table, or a computer program listing appendix submitted on a compact disc and an incorporation-by-reference of the material on the compact disc (see 37 C.F.R. § 1.52(e)(5)).

(6) Background of the invention.

(7) Brief summary of the invention.

(8) Brief description of the several views of the drawing.

(9) Detailed description of the invention.

(10) A claim or claims.

(11) Abstract of the disclosure.

(12) Sequence Listing listing in a proscribed format polynucleotide and/or polypeptide sequences listed in the application.

The patent specification is the part of the application that describes the invention in sufficient detail to teach one of skill in the art how to make and use the invention without an undue amount of experimentation. It also should describe any experiments showing that the invention performs as conceived by the inventor and any minor modifications and substitutions that can be made to the invention. The specification optionally may describe the technical background and commercial application of the invention and may distinguish the invention from other similar technologies and any unexpected benefits or results of the invention.

The patent claims are key to the enforcement of a patent right because they describe the scope of patent rights. The claims are negotiated with the USPTO during the examination process and are therefore altered or "amended" to a form that is determined by the USPTO to satisfy the criteria for patentability. Most patent applications contain several to many claims, and can describe several inventions in one patent application.

2.3.4   Fees, Surcharges, and Signed Declaration

Filing fees and other associated surcharges are also required before the USPTO will begin the examination process. In addition, before the grout of the patent, the inventor or inventors must submit a signed oath or declaration attesting to the fact that each named inventor is in fact an inventor of at least one claim of the nonprovisional application.

   Although an inventor may prepare, file and negotiate with the USPTO (a process called "prosecuting" a patent application), it is highly recommended that an experienced professional be involved with the preparation and prosecution of the patent application to minimize defects in the application. A patent professional can be an attorney or a patent agent, each of which must be licensed to represent inventors before the USPTO. For attorneys, this licensure is separate and in addition to licensure with any state bar organization.

2.3.5   Filing Receipt and Foreign Filing License

After all the appropriate papers and fees are received by the USPTO, a filing date, serial number and confirmation number will be assigned to the application. In addition, a projected publication date of the application will be assigned to the application.[8] This information is provided to the inventor on a "filing receipt" which is mailed to the correspondence address provided by the inventor. The filing receipt may also indicate that the inventor may without further authorization, send the application outside the United States for filing internationally. This is termed a

---

[8]The USTPO now publishes applications 18 months from the date of the earliest priority date unless a patent applicant qualifies for nonpublication and a petition is filed at the time the non-provisional application is filed requesting nonpublication. See 37 C.F.R. § 1.213 for the criteria for nonpublication.

"foreign filing license." If the inventor desires to send the application to a foreign jurisdiction for international filing before the receipt of the foreign filing license, a petition can be filed with the USPTO requesting authorization to export the application outside the United States.[9]

2.3.6 Paris Convention Treaty

The United States is a member of the Paris Convention Treaty ("PCT"), an international treaty that allows a U.S. inventor to file foreign patent applications on the same invention as the U.S. application and obtain the benefit of the original US filing date provided the foreign application is filed on or before the one year anniversary date of the original patent filing. In addition, one international appli-

---

[9] 35 U.S.C. § 181 states that whenever publication or disclosure by the publication of an application or by the grant of a patent on an invention in which the Government has a property interest might, in the opinion of the head of the interested Government agency, be detrimental to the national security, the Commissioner of Patents upon being so notified shall order that the invention be kept secret and shall withhold the publication of an application or the grant of a patent therefor under the conditions set forth hereinafter. Thus, the USPTO reviews patent filings to determine if the subject matter of the application would fall under section 181. After such a determination or the expiration of six months from the filing date of the application, the USPTO issues a foreign filing license thereby allowing the foreign filing of the application. International patent applications filed under the PCT, if filed within the United States Receiving Office, are exempt from obtaining a foreign filing license.

The laws relating to the filing of a patent application in a foreign country are found in 35 U.S.C. §§ 184–185. 35 U.S.C. § 184 states in part that "[e]xcept when authorized by a license obtained from the Commissioner of Patents a person shall not file or cause or authorize to be filed in any foreign country prior to six months after filing in the United States an application for patent or for the registration of a utility model, industrial design, or model in respect of an invention made in this country. A license shall not be granted with respect to an invention subject to an order issued by the Commissioner of Patents pursuant to section 181 of this title [Title 35] without the concurrence of the head of the departments and the chief officers of the agencies who caused the order to be issued. The license may be granted retroactively where an application has been filed abroad through error and without deceptive intent and the application does not disclose an invention within the scope of section 181 of this title [Title 35]."

35 U.S.C. 185 of the patent statute provides that if no person shall obtain a patent in the United States if that person filed a patent application in a foreign country without first obtaining a foreign filing license. A United States patent issued to such person, his successors, assigns, or legal representatives shall be invalid, unless the failure to procure such license was through error and without deceptive intent, and the patent does not disclose subject matter within the scope of section 181 of Title 35.

cation can be filed with a patent office of a member country (a "PCT International Application"). The timely filed international application preserves the right and ability to file nationally or regionally based applications in the member countries of the PCT at a later date. Most developed countries are members of the PCT, with the exception of Taiwan. However, Taiwan does have a separate agreement with the United States that allows reciprocity and benefit of the US application date.

This one-year "PCT" anniversary date runs concurrently with the one-year anniversary date of the refiling of a provisional application to a nonprovisional application. Thus, it is common for a provisional application to be filed with an international application filed on or before the one-year anniversary date. The USPTO is an authorized receiving authority for the filing of international applications and therefore international PCT applications can be filed in the USPTO, provided at least one inventor or applicant (i.e., an assignee of the patent application) is a United States citizen or resident.

2.3.7   Foreign Applications

Alternatively or in addition, the foreign application may be filed directly in the patent office of the country where foreign patent protection is sought. Typically, the patent agent or attorney engages an individual or law firm registered to practice before the foreign patent office to assist with the filing of the patent application. Foreign countries where the invention will be made, sold or used are typically selected for foreign filing and pursuit of patent protection.

The deadline to file a nonprovisional application (that may also be an international application) is a time to revisit the research and development of the invention and supplement the original patent filing with any new related inventions or experiments.

The USPTO also requires inventors, applicants and anyone involved with the invention and patent application to disclose to the USPTO any documents or information that is material to the patentability of the application. This material includes technical and patent literature that could be relevant as "prior art" to the claims. This information is submitted to the USPTO in a document termed an "information disclosure statement." Patent applicants are required to update the information disclosure statement with any information of which they are aware up to the time of the issuance of a patent on the filed application.

2.4   EXAMINATION OF THE CLAIM

After the USPTO has received the necessary fees and documents from the inventors, the application is assigned to a patent examiner. The patent examiner typically reviews the claims and determines whether one or more inventions have been included in the one nonprovisional filing. If the examiner determines that more than one invention has been claimed, a requirement for restriction is sent to the inventor or his attorney or agent and a request is made to elect a single invention for initial examination. The nonelected claims that embody the nonelected inventions must be pursued in a "divisional" application to be filed at a later date. The divisional applications do not lose the benefit of their filing date, but they also do not gain any additional patent term (other than additional patent term for USPTO delays or regulatory review, as appropriate).

Upon the election of claims to one invention or if no election was required, the USPTO will review the claims and patent specification for compliance with the requirements for patentability and conduct a search of the literature (technical and patent literature) for "prior art" that either negates novelty of the claims or brings into question the nonobviousness of the examined claims.

The examiner then prepares a document called an "office action" setting forth reasons why the claims and patent specification allegedly fail to satisfy the patent laws and rules. The issuance of an office action should not be interpreted that no patent can be issued on the invention but rather as a first negotiation with the inventor or applicants as to the scope of the examined claims. The granting of a patent at first pass by the USPTO examiner is very unusual and highly unlikely.[10]

The office action may list a number of technical and patent publications that the examiner considers to disclose directly (i.e. anticipate) or suggest (i.e., render the claim "obvious") the examined claims. For a reference to anticipate the rejected claim, the reference must disclose in the one document all elements or teachings of the individual claim. For a reference to render obvious a claim, the reference alone or in combination with other cited reference or information well known in the field of the invention, must suggest all elements of the individual claim.

The office action may also reject the claims for allegedly trying to encompass subject matter not supported by the specification with experiments or details. In addition, the claim terms themselves could be rejected for using terms that are "vague and indefinite" in that they fail to properly instruct one of skill in the art of the scope of the claims.

The claims could also be rejected for allegedly trying to claim prohibited subject matter. In the context of regenerative medicine, one cannot seek to patent a natural law or product of nature, e.g., a stem cell as it exists in the human body. A means to distinguish this patent-ineligible subject matter from patent-eligible subject matter is to indicate that the claimed stem cell is in isolated or purified form.

In sum, the USPTO examiner will examine the application and the claims to determine if they meet the legal requirements for patentability (novelty and non-obviousness), and whether the application meets the subject matter and description requirements.

After receipt and review of the office action, the inventor and his counsel will determine if the examiner has a valid rejection or if a misunderstanding on the part of the examiner was the basis for the rejection. The inventor's rebuttal arguments and if necessary claim amendments are submitted to the USPTO examiner

---

[10] See generally, footnote 1.

in writing within the time specified by the patent rules, i.e., by no later than six months from the date of issuance of the office action. Failure to reply within this time period will result in abandonment of the application. Any response filed within three months of the issuance of the office action can be submitted without an additional fee. Any response filed after the three month deadline will require an "extension fee."[11]

In addition to filing a paper reply to the office action, the USPTO examiner may allow the inventor and her counsel to discuss the merits of the application and the rejections in person or in a telephonic or personal interview.

It is important to note that any amendments to the claims to overcome and remove the rejections in the office action must be based on information already present in the originally filed application or on information known in the technical field of the invention at the time of the filing of the patent application. If necessary, new information (usually in the form of experimental data) can be submitted to demonstrate the properties and/or operation of the invention. This information is submitted in the form of a declaration or affidavit signed by the person who obtained the experimental data or supervised the work of obtaining the experimental data. This data and information is not incorporated into the text of the patent issuing from the application, but is placed in the USPTO examination file (a.k.a. the "file wrapper" or "prosecution history" for the patent).[12]

The examiner reviews the inventor's response to the office action and may telephone with suggested amendments to place the application in condition for allowance or may allow the application. Alternatively, the examiner may issue a second office action that may be identified as a "final office action" that may withdraw or maintain some of the rejections.

The inventor may choose to respond to the final office action with further evidence to rebut the examiner's rejection or amend the claims to move the application to allowance. If the second reply does not place the application in condi-

---

[11] See footnote 1.
[12] *Id.*

tion for allowance, the inventor may choose to appeal the application, refile the application, abandon the application and/or pursue claims that were not examined in the first case in a continuation or divisional application.[13]

If the inventor decides to pursue a continuation or divisional application, the continuation or divisional application must be filed before the patenting or abandonment of the examined application.

2.5 GRANTING OF THE PATENT

At the time the examiner determines that the claims and application satisfied the criteria for patentability, the examiner would issue a notice of allowability that will set a deadline of three months from date of issuance of the notice to pay an issue fee. This deadline cannot be extended and failure to pay the fee will result in abandonment of the application.[14]

The patent term can be adjusted for delays caused by the USPTO or by delays caused by the inventor during the examination process. This is calculated at the time of allowance and at the time of issuance of the patent. The calculation is complicated and the USPTO has a program that will assist inventors with the calculation.

The patent is printed and mailed to the correspondence address. For applications that are filed after November 29, 2000, the USPTO not only publishes the application but also makes available to the public the prosecution history of the patent. The file history can be obtained through the USPTO Public PAIR system. The Public PAIR only displays issued or published application statuses. To access Public PAIR, one needs to have a patent, application, or publication number that you wish to search. An inventor can access documents in his own patent

---

[13] *Id.*
[14] *Id.*

application through Private PAIR. To use Private PAIR, an inventor needs a digital certificate issued from the USPTO's Public Key Infrastructure.[15]

In addition, for applications filed after November 29, 2000, inventors can obtain reasonable royalties from others who make, use, sell, or import the invention during the period between the time the patent application is published and the patent is granted, provided the claims as published in the application are substantially identical to the issued claims. An inventor can also request that applications be published earlier than 18 months, a procedure that offers inventors provisional rights at an earlier stage.[16]

After the patent issues, patent maintenance fees must be paid at 3 and ½ and 7 and ½ years after issuance of the patent. Failure to pay the fees by the deadlines (or during a grace period with a surcharge) will result in abandonment of the patent rights. The USPTO maintains a database of patents that have expired for failure to timely pay a maintenance fee.

Recall, the patent claims define the scope of patent monopoly. Thus, information disclosed but not claimed in a patent (or a related patent filing like a continuation or divisional application) is not part of the patent monopoly.

· · · ·

---

[15] See uspto.gov/patents/process/search/public_pair/index.jsp, for more information on the Public PAIR and Private PAIR system.
[16] *Id.*

CHAPTER 3

Patenting Stem Cell Technologies

As noted in the previous chapters, any new and useful process, machine, article of manufacture, or composition of matter, or any new and useful improvement thereof can be patentable. In the context of regenerative medicine the following technologies are typically patented: methods for conducting business in regenerative medicine; isolated stem cells; methods for isolating, culturing, differentiating and de-differentiating stem cells; compositions to isolate or differentiate or culture the cells; populations of cells; and the use of the cells diagnostically, therapeutically, and for research purposes.

For an overview of stem cell technologies patented in the United States, The California Institute for Regenerative Medicine has published a summary of all U.S. patents relating to stem cell technologies through 2010, which is available on its web site.[1]

3.1   ISOLATED STEM CELLS

Isolated stem cells have been patented in the United States for many years. Whether or not a patent can be granted on the stem cell depends on the type or kind of stem cell (e.g., adult, embryonic or induced) and the country where patent protection is sought.

---

[1] See Resources for Stem Cell Researchers, U.S. Patent Database, available at www.cirm.ca .gov/our-funding/stem-cell-research-resources. The Hinxton Group has published an overview of the patent-eligibility of stem cells by types and country. The report "Stem Cell Research Patent Landscape (Briefing Note)" can be accessed at the website, hinxtongroup.wordpress.com /background-2/ip-landscape/.

In addition to qualifying for patent-eligibility in the country in which protection is sought, the patent claim must identify the cell or cell population as "isolated" or "purified" to distinguish the cell or population from the cells, as it exists in the human body or other animal from which it was isolated.

Early patents to adult or somatic stem cells describe the cells in terms of the tissue from which the cell was separated and/or the ability of the cell to differentiate into cells of certain lineage. U.S. Patent No. 5,486,359, issued on January 23, 1996, provides an example of an early patent claim to an isolated stem cell population:

> "1. An isolated, homogeneous population of human mesenchymal stem cells which can differentiate into cells of more than one connective tissue type."

More recently, isolated cells are described in patents by identifying markers. U.S. Patent No. 8,119,398, issued on February 21, 2012, provides an example of a recently issued claim to an isolated stem cell:

> "1. A therapeutic composition for promoting tissue regeneration in a mammal comprising isolated adipose-derived stem cell side population (ADSC-SP) cells, wherein said side population consists essentially of cells having the phenotype Lin^-, $Sca\text{-}1^+$, $CD90^+$, $CD34^{low}$, $CD13^{low}$, $CD117^-$ and $CD18^{low}$."

In addition to an isolated single cell and provided the work done by the inventor supports the claim, substantially homogenous populations of stem cell types and compositions containing the cells and/or populations also can be claimed. The cells and/or populations can be modified by the introduction of exogenous factors, such as exogenous nucleic acids and proteins.

3.2   EMBRYONIC STEM CELLS

United States law also allows the patenting of embryonic stem cells, including human embryonic stem cells (hESCs). Dr. James A. Thomson was one of the first to patent embryonic stem cells in the United States and he continues to patent such technologies. On September 25, 2012, he was awarded a patent entitled "Preparation of human embryonic stem cells." Claim 1 of the patent recites:

> "1. A preparation of pluripotent human embryonic stem cells comprising cells that (i) proliferate in vitro for over one year, (ii) maintain a karyotype in which the chromosomes are euploid through prolonged culture, (iii) maintain the potential to differentiate to derivatives of endoderm, mesoderm, and ectoderm tissues, (iv) are inhibited from differentiation when cultured on a fibroblast feeder layer, and (v) are negative for the SSEA-1 cell surface marker and positive for SSEA-4 cell surface marker."

Similar to adult stem cell populations and provided the work done by the inventor supports the claim, substantially homogenous populations of embryonic stem cell and compositions containing the cells and/or populations also can be claimed. The cells and/or populations can be modified by the introduction of exogenous factors, such as exogenous nucleic acids and proteins.

In contrast, the countries of the European Union do not allow the patenting of human embryonic stem cells or any product, which requires the destruction of human embryos, even if the patent claim does not specifically refer to the use of a human embryo. For the purpose of the exclusion, the term "human embryo" is interpreted broadly to include any organism that is capable of commencing the process of development of a human being. This includes: a human ovum as soon as fertilized, if that fertilization is such as to commence the process of development into a human being; a non-fertilized human ovum into which the cell nucleus from a mature human cell has been transplanted, insofar as it is capable of commencing the process of development of a human being; and a non-fertilized human ovum whose division and further development have been stimulated by parthenogenesis, insofar as it is capable of commencing the process of development of a human being.[2]

China is also reported to oppose the patenting of human embryonic stem cells as being contrary to morality and public interest.[3]

3.3 INDUCED PLURIPOTENT STEM CELLS (iPSCs)

Induced pluripotent stem cells (iPSCs), including human iPSCs are also patentable in the United States. Because an embryo is not destroyed in the creation of the cells or their use, human iPSCs are considered to patent-eligible in all countries.

U.S. Patent No. 7,682,828, entitled "Methods for reprogramming somatic cells" is an early patent claiming iPSCs and names coinventors Rudolf Jaenisch and Konrad Hochedlingeris assigned to the Whitehead Institute. It issued on March 23, 2010. Claim 1 of the patent recites:

> "1. A primary somatic cell comprising in its genome a first endogenous pluripotency gene operably linked to DNA encoding a first selectable marker in such a manner that expression of the first selectable marker substantially matches expression of the first endogenous pluripotency gene, wherein the cell additionally comprises an exogenously introduced nucleic acid encoding a pluripotency protein and operably linked to at least one regulatory sequence, wherein the endogenous

---

[2] See, "Inventions involving human embryonic stem cells" available at www.ipo.gov.uk/pro-types /pro-patent/p-law/p-pn/p-pn-stemcells-20120517.htm.
[3] See Zhu, "Comparative Study on Patenting of Human Embryonic Stem Cells" available at patent docs.org/2011/11/comparative-study-on-patenting-of-human-embryonic-stem-cells.html.

pluripotency gene is a gene that is expressed in a pluripotent embryonic stem cell, is required for the pluripotency of the embryonic stem cell, and is downregulated as the embryonic stem cell differentiates, and wherein the pluripotency protein is a protein expressed in a pluripotent embryonic stem cell, is required for the pluripotency of the embryonic stem cell, and is downregulated as the embryonic stem cell differentiates."

Similar to adult stem cell populations and provided the work done by the inventor supports the claim, substantially homogenous populations of iPSCs and compositions containing the cells and/or populations also can be claimed. The cells and/or populations can be modified by the introduction of exogenous factors, such as exogenous nucleic acids and proteins.

3.4   PROCESS PATENTS

Provided that the criteria for patentability are met, methods for isolating stem cells and creating them (with respect to iPSCs) are the subject matters of issued U.S. patents. In addition, compositions and methods for culturing the cells to differentiate them to certain cell types are also the subject of U.S. patent claims. An example is U.S. Patent No. 8,361,502, which issued on January 29, 2013 and is entitled "Compositions and methods for the expansion and differentiation of stem cell." Claim 1 of the patent recites:

"1. A nanofiber composition for the expansion or differentiation of stem cells comprising a core of one or more electrospun polymers, wherein an additional polymer is chemically grafted onto the one or more electrospun polymers, and said one or more electrospun polymers comprise polyethersulfone (PES)."

There are a growing number of U.S. patents that cover various methods for generating iPSCs. The patents claim the use of various reprogramming factors and conditions as well as the methods to introduce the factors into the cells for reprogramming.

An early patent to the process of preparing an iPSC issued on November 15, 2011. U.S. Patent No. 8,058,065, entitled "Oct3/4, Klf4, c-Myc and Sox2 produce induced pluripotent stem cells" issued to Dr. Shinya Yamanaka and his group at Kyoto University in Japan.[4] Only one claim issued in this patent and it recites:

"1. A method for preparing an induced pluripotent stem cell by nuclear reprogramming of a somatic cell from a mammalian species, comprising: a) introducing into the somatic cell one or more retroviral vectors comprising a gene encoding Oct3/4, a gene encoding Klf4, a gene encoding c-Myc and a gene encoding Sox2 operably linked to a promoter; and b) culturing the transduced somatic cell on a fibroblast feeder layer or extracellular matrix in a cell media that supports growth of ES cells of the mammalian species, wherein one or more pluripotent cells are obtained."

Dr. James A. Thomson and co-inventor Junying Yu also have patents related to iPSCs. The patents are assigned to Wisconsin Alumni Research Foundation and one such patent issued on September 18, 2012. This patent is entitled "OCT4 and SOX2 with SV40 T antigen produce pluripotent stem cells from primate somatic cells." Claim 1 of the patent recites:

"1. A method of reprogramming primate somatic cells, the method comprising the steps of: introducing into a primate somatic cell at

---

[4] Additional U.S. patents issued to Dr. Yamanaka and his group include U.S. Patent No. 8,278,104 and 8,129,187.

least one non-viral episomal vector that encodes at least one potency-determining factor under conditions sufficient to reprogram the cells, wherein the at least one vector encodes at least potency-determining factors OCT4 and SOX2, wherein the cells express the introduced OCT4 and SOX2 and further comprise SV40 T Antigen; and culturing the cells obtained from the introducing step in a medium that supports pluripotent cell growth to obtain pluripotent reprogrammed cells substantially free of any vector component associated with introducing the potency-determining factors to the somatic cells."

3.5   METHODS FOR USING STEM CELLS

Methods to use stem cell populations and compositions are also patentable in the United States. The cells and compositions can be used *in vitro* or *ex vivo*, in drug discovery methods or used therapeutically to treat disease or disorders. They also have been claimed as methods to introduce genes into patients as a gene therapy vector.

U.S. Patent No. 8,318,488, issued on November 27, 2012 entitled "Assay for drug discovery based on in vitro differentiated cells" is an example of a patent claiming screening methods. Claim 1 of the patent recites:

"1. An in vitro method for screening and identifying a candidate substance capable of ameliorating hypertrophic cardiomyopathy comprising: (a) differentiating a pluripotent stem cell to a cardiomyocyte in vitro; (b) inducing the cardiomyocyte to display a first cardiac hypertrophic phenotype; (i) wherein the inducing results in the cardiomyocyte having an increased size, protein production and/or sarcomeric assembly relative to the same cardiomyocyte prior to the inducing in step (b); and/or (ii) wherein the inducing results in the cardiomyocyte expressing atrial natriuretic factor (ANF) and/or brain natriuretic protein (BNP) genes or their products in an amount greater than the cardiomyocyte prior to the inducing in step (b); (c) contacting a test sample comprising the in vitro differentiated cardiomyocyte with a substance to be tested prior, during or after said cardiomyocyte is induced to display the cardiac hypertrophic phenotype; (d) measuring a second cardiac hypertrophic phenotype in the cardiomyocyte of step (c); (e) comparing the measurement obtained in step (d) to that of a cardiomyocyte not subjected to the substance to be tested; wherein the candidate substance is identified on the basis of whether the measurement of the second cardiac hypertrophic phenotype in the cardiomyocyte of step (d) is consistent with a reduction in hypertrophic cardiomyopathy."

U.S. Patent No. 7,875,273 issued on January 25, 2011 and entitled "Treatment of Parkinson's disease and related disorders using postpartum derived cells" is an example of a patent claiming treatment of disease using stem cells. Claim 1 of the patent recites:

"1. A method of treating a patient having Parkinson's Disease or parkinsonism, the method comprising administering to striatum of the patient cells in an amount effective to treat Parkinson's Disease or parkinsonism, wherein the cells are isolated from mammalian umbilical cord tissue substantially free of blood or are expanded in culture from a cell isolated from mammalian umbilical cord tissue substantially free of blood, wherein the cells are capable of self-renewal and expansion in culture, and require L-valine for growth; and wherein the cells do not produce CD117."

Some countries do not allow the patenting of medical methods to treat humans, e.g., European countries and China. However, the compositions used in the medical methods are patentable. Thus, it is advisable to contact a patent attorney or agent with knowledge of patenting in these jurisdictions when considering patenting therapeutic methods outside the United States.

*   *   *   *

CHAPTER 4

Conclusion

Many stem cell technologies and inventions related to embryonic, adult, and induced pluripotent stem cells and their use are patented in the United States provided that the inventions satisfy the general criteria for patentability: novelty, non-obviousness, and patent-eligibility and the patent application describes the invention with particularity and detail. The patenting process begins with the invention and a U.S. patent application must be filed within one year of the inventor's disclosure or public use or publication of the invention. A patent is a business tool that can be used to prevent others from using a patented invention during the term of the patent and as such, is legal monopoly for a limited time period. The monopoly can be used to allow the patent holder the time to commercialize the claimed invention.

·　·　·　·

TITLES OF RELATED INTEREST

Colloquium Series on
Stem Cell Biology

Wenbin Deng, Ph.D., *Department of Cell Biology and Human Anatomy, Institute for Pediatric Regenerative Medicine, School of Medicine, University of California, Davis*

Published Titles

Stem Cells and Extracellular Matrices
Lakshmi Chelluri
Global Hospitals, Lakdi-ka-Pool, India

Biobanking in the Era of the Stem Cell: A Technical and Operational Guide
Jennifer C. Moore, Michael H. Sheldon, Ronald P. Hart
NIMH Stem Cell Center, Rutgers University

Stem Cells and Progenitors in Liver Development
Marcus O. Muench
Blood Systems Research Institute

Forthcoming Titles

The Biology and Therapeutic Potential of Stem Cells of the Oral Cavity
Sandu Pitaru
Tel Aviv University

Emerging Roles for Neural Stem Cells in Memory, with an Emphasis on Emotional Memory
Daniela Kaufer and Aaron Friedman
University of California, Berkeley

Homing of Stem Cells to Ischemic Myocardium
Jon C. George
Temple University School of Medicine

Immunogenicity of Induced Pluripotent Stem Cells
Yang Xu
University of California, San Diego

Neural Stem Cells in Hypoxic-Ischemic and Hemorrhagic Brain Injury in Newborns
Chia-Yi Kuan
Cincinnati Children's Hospital Medical Center

Neuronal Fate Specification of Pluripotent Stem Cells
Meng Li
MRC Clinical Sciences Centre, Imperial College London (UK)

Signaling Mechanisms Controlling CNS Migration of Mesenchymal Stem Cells
Min Zhao
University of California, Davis

Stem Cell-based Therapy For Myelin Disorders
Sangita Biswas
Shriners Institute of Pediatric Regenerative Medicine

Stem Cell-based Therapy for Spinal Cord Injury
Ying Liu
Scripps Institute

Systems Biology of Neural Stem Cells: Lessons from the Olfactory Epithelium
Anne L. Calof
University of California, Irvine

For a full list of published and forthcoming titles:
http://www.morganclaypool.com/page/scb

SERIES OF RELATED INTEREST

Colloquium Series on
The Building Blocks of the Cell:
Cell Structure and Function

Editor

Ivan Robert Nabi, *Professor, University of British Columbia, Department of Cellular and Physiological Sciences*

This Series is a comprehensive, in-depth review of the key elements of cell biology including 14 different categories, such as Organelles, Signaling, and Adhesion. All important elements and interactions of the cell will be covered, giving the reader a comprehensive, accessible, authoritative overview of cell biology. All authors are internationally renowned experts in their area.

For a full list of published and forthcoming titles:
http://www.morganclaypool.com/page/bbc

Colloquium Series on
The Cell Biology of Medicine

Editors
Philip L. Leopold, PhD, *Professor and Director, Department of Chemistry, Chemical Biology, & Biomedical Engineering, Stevens Institute of Technology*
Joel Pardee, Ph.D. *President, Neural Essence; formerly Associate Professor and Dean of Graduate Research, Weill Cornell School of Medicine*

In order to learn we must be able to remember, and in the world of science and medicine we remember what we envision, not what we hear. It is with this essential precept in mind that we offer the Cell Biology of Medicine series. Each book is written by faculty accomplished in teaching the scientific basis of disease to both graduate and medical students. In this modern age it has become abundantly clear that everyone is vastly interested in how our bodies work and what has gone wrong in disease. It is likewise evident that the only way to understand medicine is to engrave in our mind's eye a clear vision of the biological processes that give us the gift of life. In these lectures, we are dedicated to holding up for the viewer an insight into the biology behind the body. Each lecture demonstrates cell, tissue and organ function in health and disease. And it does so in a visually striking style. Left to its own devices, the mind will quite naturally remember the pictures. Enjoy the show.

For a list of published and forthcoming titles:
http://www.morganclaypool.com/toc/cbm/1/1

Colloquium Series on Developmental Biology

Editors

Jean-Pierre Saint-Jeannet, Ph.D., *Professor, Department of Basic Science & Craniofacial Biology, College of Dentistry, New York University*

Daniel S. Kessler, Ph.D., *Associate Professor of Cell and Developmental Biology, Chair, Developmental, Stem Cell and Regenerative Biology Program of CAMB, University of Pennsylvania School of Medicine*

Developmental biology is in a period of extraordinary discovery and research. This field will have a broad impact on the biomedical sciences in the coming decades. Developmental Biology is interdisciplinary and involves the application of techniques and concepts from genetics, molecular biology, biochemistry, cell biology, and embryology to attack and understand complex developmental mechanisms in plants and animals, from fertilization to aging. Many of the same genes that regulate developmental processes underlie human regulatory gene disorders such as cancer and serve as the genetic basis of common human birth defects. An understanding of fundamental mechanisms of development is providing a basis for the design of gene and cellular therapies for the treatment of many human diseases. Of particular interest is the identification and study of stem cell populations, both natural and induced, which is opening new avenues of research in development, disease, and regenerative medicine. This eBook series is dedicated to providing mechanistic and conceptual insight into the broad field of Developmental Biology. Each eBook is intended to be of value to students, scientists and clinicians in the biomedical sciences.

For a full list of published and forthcoming titles:
http://www.morganclaypool.com/toc/deb/1/1

Colloquium Series on
The Genetic Basis of Human Disease

Editor

Michael Dean, Ph.D., *Head, Human Genetics Section, Senior Investigator, Laboratory of Experimental Immunology National Cancer Institute (at Frederick)*

This series will explore the genetic basis of human disease, documenting the molecular basis for rare and common Mendelian and complex conditions. The series will overview the fundamental principles in understanding such as Mendel's laws of inheritance, and genetic mapping through modern examples. In addition current methods (GWAS, genome sequencing) and hot topics (epigenetics, imprinting) will be introduced through examples of specific diseases.

For a full list of published and forthcoming titles:
http://www.morganclaypool.com/page/gbhd

Colloquium Series on
Genomic and Molecular Medicine

Editor

Professor Dhavendra Kumar, MD, FRCP, FRCPCH, FACMG, *Consultant in Clinical Genetics, All Wales Medical Genetics Service Genomic Policy Unit, The University of Glamorgan, UK Institute of Medical Genetics, Cardiff University School of Medicine, University Hospital of Wales*

From 1970 onwards, there has been a continuous and growing recognition of the molecular basis of medical practice. Alongside the developments and progress in molecular medicine, new and rapid discoveries in genetics have led to an entirely new approach to the practice of clinical medicine. Until recently the field of genetic medicine has largely been restricted to the diagnosis of disease, offering explanation and assistance to patients and clinicians in dealing with a number of relatively uncommon inherited disorders. However, since the completion of the human genome in 2003 and several other genomes, there is now a plethora of information available that has attracted the attention of molecular biologists and allied researchers. A new biological science of Genomics is now with us, with far reaching dimensions and applications.

During the last decade, rapid progress has been made in new genome-level diagnostic and prognostic laboratory methods, and revealing findings in genomics have led to changes in our understanding of fundamental concepts in cell and molecular biology. It may well be that evolutionary and morbid changes at the genome level could be the basis of normal human variation and disease. Applications of individual genomic information in clinical medicine have led to the prospect of robust evidence-based personalized medicine, and genomics has led to the discovery and development of a number of new drugs with far reaching implications in pharmacotherapeutics. The existence of Genomic Medicine around us is inseparable from molecular medicine, and it contains tremendous implications for the future of clinical medicine.

For a full list of published and forthcoming titles:
http://www.morganclaypool.com/toc/gmm/1/1

Colloquium Series on
Integrated Systems Physiology:
From Molecule to Function to Disease

Editors
D. Neil Granger, Ph.D., *Boyd Professor and Head of the Department of Molecular and Cellular Physiology at the LSU Health Sciences Center, Shreveport*
Joey P. Granger, Ph.D., *Billy S. Guyton Distinguished Professor, Professor of Physiology and Medicine, Director of the Center for Excellence in Cardiovascular-Renal Research, and Dean of the School of Graduate Studies in the Health Sciences at the University of Mississippi Medical Center*

Physiology is a scientific discipline devoted to understanding the functions of the body. It addresses function at multiple levels, including molecular, cellular, organ, and system. An appreciation of the processes that occur at each level is necessary to understand function in health and the dysfunction associated with disease. Homeostasis and integration are fundamental principles of physiology that account for the relative constancy of organ processes and bodily function even in the face of substantial environmental changes. This constancy results from integrative, cooperative interactions of chemical and electrical signaling processes within and between cells, organs and systems. This eBook series on the broad field of physiology covers the major organ systems from an integrative perspective that addresses the molecular and cellular processes that contribute to homeostasis. Material on pathophysiology is also included throughout the eBooks. The state-of the art treatises were produced by leading experts in the field of physiology. Each eBook includes stand-alone information and is intended to be of value to students, scientists, and clinicians in the biomedical sciences. Since physiological concepts are an ever-changing work-in-progress, each contributor will have the opportunity to make periodic updates of the covered material.

For a full list of published and forthcoming titles:
http://www.morganclaypool.com/toc/isp/1/1

Colloquium Series on Neurobiology of Alzheimer's Disease

Editors

George Perry, Ph.D., *Professor of Biology and Dean of the College of Sciences, University of Texas, San Antonio*

Rudolph J. Castellani, M.D., *Professor, Pathology, University of Maryland, School of Medicine*

This e-book series on Alzheimer's disease will provide an up-to-date, comprehensive overview of Alzheimer's disease and dementia from a multidisciplinary perspective, with a focus on disease pathogenesis and translational neurobiology. The major pathogenic proteins will be described and discussed in depth from the perspective of molecular and cell biology, experimental and transgenic modeling, and in-situ phenotypic expression in humans. Added to this will be in depth discussions of all the major pathogenic theories, including the amyloid and tau protein cascades, oxidative stress, involvement of heavy metals, interplay of the endocrine system, issues surrounded cell cycle activation and protein signaling, and important comorbidities that influence human disease such as vascular neurobiology and synucleinopathy. Treatment paradigms and trials as a function of known components of disease pathogenesis will also be described and discussed in requisite detail that reflects the state of the art. As such, the eBook series will provide a "one stop shop" for the aspiring neuroscientist or physician scientist, and will prove an adaptable framework that can be updated going forward, as dictated by new discoveries and the accumulating scientific literature.

For a list of published and forthcoming titles:
http://www.morganclaypool.com/page/alz

Colloquium Series on Neuropeptides

Editors

Lakshmi Devi, Ph.D., *Professor, Department of Pharmacology and Systems Therapeutics, Associate Dean for Academic Enhancement and Mentoring, Mount Sinai School of Medicine, New York*

Lloyd D. Fricker, Ph.D., *Professor, Department of Molecular Pharmacology, Department of Neuroscience, Albert Einstein College of Medicine, New York*

Communication between cells is essential in all multicellular organisms, and even in many unicellular organisms. A variety of molecules are used for cell-cell signaling, including small molecules, proteins, and peptides. The term 'neuropeptide' refers specifically to peptides that function as neurotransmitters, and includes some peptides that also function in the endocrine system as peptide hormones. Neuropeptides represent the largest group of neurotransmitters, with hundreds of biologically active peptides and dozens of neuropeptide receptors known in mammalian systems, and many more peptides and receptors identified in invertebrate systems. In addition, a large number of peptides have been identified but not yet characterized in terms of function. The known functions of neuropeptides include a variety of physiological and behavioral processes such as feeding and body weight regulation, reproduction, anxiety, depression, pain, reward pathways, social behavior, and memory. This series will present the various neuropeptide systems and other aspects of neuropeptides (such as peptide biosynthesis), with individual volumes contributed by experts in the field.

For a list of published and forthcoming titles:
http://www.morganclaypool.com/toc/npe/1/1

Colloquium Series on Neuroglia
From Physiology to Disease

Editors

Alexej Verkhratsky, Ph.D., Professor of Neurophysiology, University of Manchester

Vlad Parpura, Ph.D., Associate Professor of Neurophysiology, University of Alabama at Birmingham

For decades the neuroglia were known, and known to be numerous, in the neural system. Nevertheless they were long thought to play only a minor, supporting role to other cells such as axons and neurons. Glial cell are now recognized are essential to neural functioning and represent an exciting, rapidly growing field in the neurosciences. This series will explore the overall molecular physiology of glial cells as well as their role in pathologic conditions.

For a full list of published and forthcoming titles:
http://www.morganclaypool.com/page/neuroglia

Colloquium Series on
Protein Activation and Cancer

Editors

A. Majid Khatib, Ph.D. *Research Director, INSERM, and, University of Bordeaux*

This series is designed to summarize all aspects of protein maturation by proprotein convertases in cancer. Topics included deal with the importance of these processes in the acquisition of malignant phenotypes by tumor cells, induction of tumor growth, and metastasis. This series also provides the latest knowledge on the clinical significance of convertase expression and activity, and the maturation of their protein substrates in various cancers. The potential use of their inhibition as a therapeutic approach is also explored.

For a full list of published and forthcoming titles:
http://www.morganclaypool.com/page/pac/1/1